Preface

Welcome to the Science Olympiad All Chapters Combined MCQ Class 9!

This book is specially crafted to ignite the spark of curiosity and foster a love for science in young minds. Designed with the unique needs of first graders in mind, it provides a comprehensive introduction to fundamental scientific concepts through engaging, age-appropriate content.

Our primary goal is to make learning science an enjoyable and enriching experience. The book is filled with colorful illustrations, real-life examples, and interactive exercises that help children understand and relate to the world around them. Each chapter is carefully structured to build on prior knowledge, ensuring a steady progression in learning.

Key features of this book include:

Interactive Exercises: Hands-on activities and simple experiments designed to engage students actively and make learning fun.

Multiple Choice Questions: End-of-chapter MCQs help reinforce learning and provide ample practice for young learners.

Cross-Curricular Connections: Integrates elements of math and language arts, promoting a holistic learning experience.

Environmental Awareness: Introduction to basic environmental science and sustainability concepts.

Through this book, we aim to encourage critical thinking, problem-solving skills, and a deeper understanding of scientific principles. The inclusion of real-life scientists and their contributions will inspire young learners and highlight the importance of science in our daily lives.

Whether used in the classroom or at home, this book is an invaluable resource for fostering a lifelong love of science. We hope that it will be a stepping stone for students, helping them explore and appreciate the wonders of the scientific world.

Thank you for choosing our Science Olympiad Book. Let's embark on this fascinating journey of discovery and learning together!

Happy Learning!

Table of Contents

Questions: Part 1

1. Which of the following is not a characteristic of particles of matter?

   a) They are very large

   b) They attract each other

   c) They are in continuous motion

   d) They have space between them

2. Which state of matter has a definite shape and a definite volume?

   a) Solid

   b) Liquid

   c) Gas

   d) Plasma

3. The smallest particle of an element that can take part in a chemical reaction is:

   a) Molecule

   b) Atom

   c) Proton

   d) Neutron

4. The cell wall in plants is mainly composed of:

   a) Cellulose

   b) Protein

   c) Lipid

   d) Starch

5. Work done is equal to:

a) Force × Distance

b) Mass × Acceleration

c) Mass × Velocity

d) Force × Time

6. Which law states that 'to every action, there is an equal and opposite reaction'?

a) Newton's First Law

b) Newton's Second Law

c) Newton's Third Law

d) Law of Conservation of Energy

7. Which of the following is a vector quantity?

a) Distance

b) Speed

c) Displacement

d) Time

8. The value of acceleration due to gravity (g) on Earth is approximately:

a) 9.8 m/s^2

b) 8.9 m/s^2

c) 10.8 m/s^2

d) 11.2 m/s^2

9. Which of the following is a pure substance?

a) Air

b) Milk

c) Gold

d) Sand

10. The nucleus of an atom contains:

a) Protons only

b) Neutrons only

c) Protons and neutrons

d) Electrons and protons

11. Which type of tissue is responsible for movement in our body?

a) Epithelial tissue

b) Connective tissue

c) Muscle tissue

d) Nervous tissue

12. The scientific name of mango is:

a) Mangifera indica

b) Panthera leo

c) Rosa indica

d) Oryza sativa

13. The rate of change of displacement is called:

a) Speed

b) Velocity

c) Acceleration

d) Momentum

14. Xylem and phloem are examples of:

a) Muscular tissues

b) Nervous tissues

c) Epithelial tissues

d) Vascular tissues

15. The unit of force in the SI system is:

a) Newton

b) Joule

c) Pascal

d) Watt

16. Which of the following is a method of increasing soil fertility?

a) Ploughing

b) Irrigation

c) Crop rotation

d) Sowing

17. Which gas is primarily responsible for the greenhouse effect?

a) Oxygen

b) Nitrogen

c) Carbon dioxide

d) Hydrogen

18. The unit of frequency is:

a) Hertz

b) Meter

c) Second

d) Decibel

19. Which of the following is a bacterial disease?

a) Malaria

b) AIDS

c) Tuberculosis

d) Ringworm

20. The force of attraction between any two bodies is called:

a) Electrostatic force

b) Magnetic force

c) Gravitational force

d) Nuclear force

21. The smallest particle of an element that can take part in a chemical reaction is:

a) Molecule

b) Atom

c) Proton

d) Neutron

22. The nucleus of an atom contains:

a) Protons only

b) Neutrons only

c) Protons and neutrons

d) Electrons and protons

23. Which of the following has the smallest number of molecules?

a) 1g of H_2

b) 1g of O_2

c) 1g of CO_2

d) 1g of N_2

24. The maximum number of electrons in the second shell of an atom is:

a) 2

b) 8

c) 18

d) 32

25. The cell wall in plants is mainly composed of:

a) Cellulose

b) Protein

c) Lipid

d) Starch

26. Which organelle is known as the 'powerhouse of the cell'?

a) Nucleus

b) Ribosome

c) Mitochondria

d) Chloroplast

27. The main source of energy for the Earth is:

a) Wind

b) Sun

c) Water

d) Fossil fuels

28. The term 'pisciculture' refers to:

a) Cultivation of fish

b) Cultivation of crops

c) Cultivation of fruits

d) Cultivation of vegetables

29. Sound is a form of:

a) Chemical energy

b) Light energy

c) Mechanical energy

d) Nuclear energy

30. The rate of change of displacement is called:

a) Speed

b) Velocity

c) Acceleration

d) Momentum

31. The value of acceleration due to gravity (g) on Earth is approximately:

a) 9.8 m/s^2

b) 8.9 m/s^2

c) 10.8 m/s^2

d) 11.2 m/s^2

32. Diseases that can be transmitted from one person to another are called:

a) Non-communicable diseases

b) Communicable diseases

c) Chronic diseases

d) Acute diseases

33. Which of the following is a bacterial disease?

a) Malaria

b) AIDS

c) Tuberculosis

d) Ringworm

34. The force of attraction between any two bodies is called:

a) Electrostatic force

b) Magnetic force

c) Gravitational force

d) Nuclear force

35. Which type of tissue is responsible for movement in our body?

a) Epithelial tissue

b) Connective tissue

c) Muscle tissue

d) Nervous tissue

36. Which kingdom includes unicellular eukaryotic organisms?

a) Monera

b) Protista

c) Fungi

d) Plantae

37. Which of the following is a method of increasing soil fertility?

a) Ploughing

b) Irrigation

c) Crop rotation

d) Sowing

38. Which gas is primarily responsible for the greenhouse effect?

a) Oxygen

b) Nitrogen

c) Carbon dioxide

d) Hydrogen

39. The unit of force in the SI system is:

a) Newton

b) Joule

c) Pascal

d) Watt

40. Which of the following is a vector quantity?

a) Distance

b) Speed

c) Displacement

d) Time

41. Work done is equal to:

a) Force × Distance

b) Mass × Acceleration

c) Mass × Velocity

d) Force × Time

42. The smallest particle of an element that can take part in a chemical reaction is:

a) Molecule

b) Atom

c) Proton

d) Neutron

43. The nucleus of an atom contains:

a) Protons only

b) Neutrons only

c) Protons and neutrons

d) Electrons and protons

44. The maximum number of electrons in the second shell of an atom is:

a) 2

b) 8

c) 18

d) 32

45. The cell wall in plants is mainly composed of:

a) Cellulose

b) Protein

c) Lipid

d) Starch

46. The main source of energy for the Earth is:

a) Wind

b) Sun

c) Water

d) Fossil fuels

47. The value of acceleration due to gravity (g) on Earth is approximately:

a) 9.8 m/s^2

b) 8.9 m/s^2

c) 10.8 m/s^2

d) 11.2 m/s^2

48. Sound is a form of:

a) Chemical energy

b) Light energy

c) Mechanical energy

d) Nuclear energy

49. Diseases that can be transmitted from one person to another are called:

a) Non-communicable diseases

b) Communicable diseases

c) Chronic diseases

d) Acute diseases

50. The rate of change of displacement is called:

a) Speed

b) Velocity

c) Acceleration

d) Momentum

Question: Part 2

51. Which kingdom includes unicellular eukaryotic organisms?

a) Monera

b) Protista

c) Fungi

d) Plantae

52. The smallest particle of an element that can take part in a chemical reaction is:

a) Molecule

b) Atom

c) Proton

d) Neutron

53. Which of the following has the smallest number of molecules?

a) 1g of H_2

b) 1g of O_2

c) 1g of CO_2

d) 1g of N_2

54. Which type of tissue is responsible for movement in our body?

a) Epithelial tissue

b) Connective tissue

c) Muscle tissue

d) Nervous tissue

55. Xylem and phloem are examples of:

a) Muscular tissues

b) Nervous tissues

c) Epithelial tissues

d) Vascular tissues

56. The nucleus of an atom contains:

a) Protons only

b) Neutrons only

c) Protons and neutrons

d) Electrons and protons

57. The cell wall in plants is mainly composed of:

a) Cellulose

b) Protein

c) Lipid

d) Starch

58. The scientific name of mango is:

a) Mangifera indica

b) Panthera leo

c) Rosa indica

d) Oryza sativa

59. The smallest particle of an element that can take part in a chemical reaction is:

a) Molecule

b) Atom

c) Proton

d) Neutron

60. Work done is equal to:

a) Force × Distance

b) Mass × Acceleration

c) Mass × Velocity

d) Force × Time

61. The term 'pisciculture' refers to:

a) Cultivation of fish

b) Cultivation of crops

c) Cultivation of fruits

d) Cultivation of vegetables

62. The unit of force in the SI system is:

a) Newton

b) Joule

c) Pascal

d) Watt

63. The rate of change of displacement is called:

a) Speed

b) Velocity

c) Acceleration

d) Momentum

64. Which of the following is a vector quantity?

a) Distance

b) Speed

c) Displacement

d) Time

65. The value of acceleration due to gravity (g) on Earth is approximately:

a) 9.8 m/s^2

b) 8.9 m/s^2

c) 10.8 m/s^2

d) 11.2 m/s^2

66. Which kingdom includes unicellular eukaryotic organisms?

a) Monera

b) Protista

c) Fungi

d) Plantae

67. Which gas is primarily responsible for the greenhouse effect?

a) Oxygen

b) Nitrogen

c) Carbon dioxide

d) Hydrogen

68. Diseases that can be transmitted from one person to another are called:

a) Non-communicable diseases

b) Communicable diseases

c) Chronic diseases

d) Acute diseases

69. Which of the following is a bacterial disease?

a) Malaria

b) AIDS

c) Tuberculosis

d) Ringworm

70. The smallest particle of an element that can take part in a chemical reaction is:

a) Molecule

b) Atom

c) Proton

d) Neutron

71. The force of attraction between any two bodies is called:

a) Electrostatic force

b) Magnetic force

c) Gravitational force

d) Nuclear force

72. The main source of energy for the Earth is:

a) Wind

b) Sun

c) Water

d) Fossil fuels

73. Sound is a form of:

a) Chemical energy

b) Light energy

c) Mechanical energy

d) Nuclear energy

74. The unit of frequency is:

a) Hertz

b) Meter

c) Second

d) Decibel

75. Which of the following is a pure substance?

a) Air

b) Milk

c) Gold

d) Sand

76. The rate of change of displacement is called:

a) Speed

b) Velocity

c) Acceleration

d) Momentum

77. The unit of force in the SI system is:

a) Newton

b) Joule

c) Pascal

d) Watt

78. The process by which a solid changes directly into a gas is called:

a) Sublimation

b) Evaporation

c) Condensation

d) Melting

79. A solution that cannot dissolve any more solute at a given temperature is called:

a) Saturated solution

b) Unsaturated solution

c) Supersaturated solution

d) Dilute solution

80. The law of conservation of mass was established by:

a) Dalton

b) Lavoisier

c) Rutherford

d) Bohr

81. The atomic number of an element is determined by the number of:

a) Protons

b) Neutrons

c) Electrons

d) Nuclei

82. The basic structural and functional unit of life is:

a) Tissue

b) Organ

c) Cell

d) Organism

83. Which tissue in plants is responsible for transporting water and nutrients?

a) Phloem

b) Xylem

c) Parenchyma

d) Collenchyma

84. The system of classifying living organisms into five kingdoms was proposed by:

a) Linnaeus

b) Whittaker

c) Darwin

d) Aristotle

85. The SI unit of velocity is:

a) Meter

b) Meter/second

c) Kilometer/hour

d) Mile/hour

86. Which of the following is an example of a non-contact force?

a) Frictional force

b) Tension

c) Gravitational force

d) Normal force

87. The acceleration of an object moving under the influence of gravity alone is known as:

a) Gravitational acceleration

b) Terminal velocity

c) Free fall

d) Projectile motion

88. The potential energy of an object is due to its:

a) Motion

b) Position

c) Mass

d) Velocity

89. Which of the following properties of sound determines its pitch?

a) Amplitude

b) Frequency

c) Speed

d) Wavelength

90. Which of the following is a viral disease?

a) Tuberculosis

b) Cholera

c) Influenza

d) Malaria

91. Which natural resource is considered renewable?

a) Coal

b) Petroleum

c) Solar energy

d) Natural gas

92. The primary nutrient added to soil to increase its fertility is:

a) Nitrogen

b) Carbon dioxide

c) Oxygen

d) Argon

93. Which method is used to separate two immiscible liquids?

a) Filtration

b) Distillation

c) Decantation

d) Evaporation

94. The smallest unit of a compound that retains its chemical properties is:

a) Atom

b) Ion

c) Molecule

d) Element

95. Who discovered the electron?

a) J.J. Thomson

b) Rutherford

c) Niels Bohr

d) James Chadwick

96. In which part of the cell does photosynthesis occur?

a) Mitochondria

b) Chloroplast

c) Ribosome

d) Nucleus

97. Which type of animal tissue is involved in the protection of organs?

a) Epithelial tissue

b) Muscle tissue

c) Nervous tissue

d) Connective tissue

98. Which classification system is based on the hierarchical organization of species?

a) Phylogenetic

b) Cladistic

c) Linnaean

d) Phenetic

99. The motion of a body in a straight line is called:

a) Curvilinear motion

b) Rectilinear motion

c) Rotational motion

d) Oscillatory motion

100. Newton's First Law is also known as the law of:

a) Inertia

b) Action and reaction

c) Force and acceleration

d) Conservation of energy

Questions: Part 3

101. The formula for calculating gravitational force between two masses is given by:

a) $F = G(m1m2/r^2)$

b) $F = ma$

c) $F = qE$

d) $F = kq1q2/r^2$

102. The ability to do work is defined as:

a) Power

b) Energy

c) Force

d) Momentum

103. Which property of sound is measured in decibels?

a) Frequency

b) Loudness

c) Pitch

d) Speed

104. A vector quantity has both:

a) Magnitude and direction

b) Speed and distance

c) Mass and volume

d) Time and velocity

105. A substance that speeds up a chemical reaction without being consumed is called a:

a) Reactant

b) Product

c) Catalyst

d) Solvent

106. The structure of an atom is best described by:

a) Bohr model

b) Dalton's model

c) Plum pudding model

d) Thomson model

107. Which component of blood is responsible for clotting?

a) Red blood cells

b) White blood cells

c) Platelets

d) Plasma

108. The biodiversity of an area is determined by:

a) The variety of species

b) The size of the ecosystem

c) The climate

d) The soil type

109. The change in position of an object over time is known as:

a) Acceleration

b) Velocity

c) Displacement

d) Speed

110. Inertia is the tendency of an object to:

a) Change its state of motion

b) Resist changes in motion

c) Increase its speed

d) Decrease its mass

111. The weight of an object is the force of:

a) Gravity acting on it

b) Friction acting on it

c) Air resistance acting on it

d) Magnetism acting on it

112. Kinetic energy depends on which two factors?

a) Mass and velocity

b) Mass and height

c) Force and distance

d) Velocity and acceleration

113. Which of the following waves are longitudinal?

a) Sound waves

b) Light waves

c) Radio waves

d) Microwaves

114. Malaria is transmitted to humans by:

a) Mosquitoes

b) Bacteria

c) Viruses

d) Parasites

115. The water cycle involves the processes of:

a) Evaporation, condensation, and precipitation

b) Photosynthesis, respiration, and decomposition

c) Melting, freezing, and sublimation

d) Erosion, weathering, and deposition

116. Which element is essential for plant growth and is often found in fertilizers?

a) Calcium

b) Phosphorus

c) Sodium

d) Magnesium

117. The atomic mass of an element is the sum of:

a) Protons and neutrons

b) Electrons and protons

c) Neutrons and electrons

d) Protons and electrons

118. Which of the following is a chemical change?

a) Melting of ice

b) Rusting of iron

c) Tearing of paper

d) Dissolving sugar in water

119. A charged atom is called:

a) Molecule

b) Ion

c) Isotope

d) Compound

120. The energy stored in an object due to its position is called:

a) Kinetic energy

b) Potential energy

c) Thermal energy

d) Chemical energy

121. Which part of the cell contains genetic material?

a) Mitochondria

b) Nucleus

c) Ribosome

d) Cytoplasm

122. Which type of tissue is responsible for transmitting signals in the body?

a) Muscle tissue

b) Nervous tissue

c) Epithelial tissue

d) Connective tissue

123. The process by which green plants make their food is called:

a) Respiration

b) Photosynthesis

c) Fermentation

d) Decomposition

124. An object that is not in motion has:

a) Kinetic energy

b) Potential energy

c) Thermal energy

d) Magnetic energy

125. The speed of sound is fastest in:

a) Air

b) Water

c) Vacuum

d) Steel

126. Which of the following substances is a compound?

a) Sodium

b) Water

c) Iron

d) Oxygen

127. Which force is responsible for holding the nucleus of an atom together?

a) Gravitational force

b) Electromagnetic force

c) Strong nuclear force

d) Weak nuclear force

128. What is the main function of the ribosome in a cell?

a) Protein synthesis

b) Lipid synthesis

c) Carbohydrate metabolism

d) DNA replication

129. Which of the following processes involves the movement of water molecules through a semi-permeable membrane?

a) Diffusion

b) Osmosis

c) Active transport

d) Endocytosis

130. The force that opposes the relative motion of two surfaces in contact is called:

a) Tension

b) Normal force

c) Friction

d) Gravitational force

131. Which of the following elements is most abundant in the Earth's crust?

a) Silicon

b) Oxygen

c) Aluminum

d) Iron

132. The bond formed by the sharing of a pair of electrons between two atoms is called:

a) Ionic bond

b) Covalent bond

c) Hydrogen bond

d) Metallic bond

133. What is the pH of a neutral solution at 25°C?

a) 0

b) 7

c) 14

d) 1

134. Which of the following is an example of an exothermic reaction?

a) Melting of ice

b) Evaporation of water

c) Combustion of methane

d) Photosynthesis

135. The energy required to remove an electron from a gaseous atom in its ground state is known as:

a) Electron affinity

b) Ionization energy

c) Electronegativity

d) Lattice energy

136. The S.I. unit of pressure is:

a) Pascal

b) Joule

c) Newton

d) Watt

137. Which of the following structures is present in both plant and animal cells?

a) Cell wall

b) Chloroplast

c) Mitochondria

d) Large central vacuole

138. The ability of a material to be stretched into a wire is known as:

a) Malleability

b) Ductility

c) Brittleness

d) Elasticity

139. Which of the following organisms is considered the "father of genetics"?

a) Charles Darwin

b) Gregor Mendel

c) Louis Pasteur

d) Robert Hooke

140. The chemical formula of sulfuric acid is:

a) HNO_3

b) H_2SO_4

c) HCl

d) H_3PO_4

141. The resistance of a conductor is directly proportional to its:

a) Length

b) Cross-sectional area

c) Temperature

d) Voltage

142. Which of the following is a primary greenhouse gas?

a) Ozone

b) Nitrous oxide

c) Chlorofluorocarbons (CFCs)

d) Carbon dioxide

143. The atomic model that introduced the concept of electron shells is known as:

a) Thomson model

b) Rutherford model

c) Bohr model

d) Quantum mechanical model

144. What is the primary function of white blood cells?

a) Transport oxygen

b) Fight infections

c) Clot blood

d) Carry nutrients

145. The process of splitting water molecules during photosynthesis is called:

a) Photolysis

b) Hydrolysis

c) Glycolysis

d) Fermentation

146. Which of the following is not a renewable energy source?

a) Solar energy

b) Wind energy

c) Natural gas

d) Hydroelectric power

147. The basic unit of a protein is:

a) Monosaccharide

b) Nucleotide

c) Amino acid

d) Fatty acid

148. Which gas is commonly used in light bulbs to prevent the filament from burning?

a) Helium

b) Argon

c) Oxygen

d) Nitrogen

149. Which part of the brain is responsible for coordinating muscle movements?

a) Cerebrum

b) Cerebellum

c) Medulla oblongata

d) Hypothalamus

150. What type of reaction occurs when an acid and a base combine to form water and a salt?

a) Oxidation

b) Reduction

c) Neutralization

d) Decomposition

Answers

1. a) They are very large

2. a) Solid

3. b) Atom

4. a) Cellulose

5. a) Force × Distance

6. c) Newton's Third Law

7. c) Displacement

8. a) 9.8 m/s^2

9. c) Gold

10. c) Protons and neutrons

11. c) Muscle tissue

12. a) Mangifera indica

13. b) Velocity

14. d) Vascular tissues

15. a) Newton

16. c) Crop rotation

17. c) Carbon dioxide

18. a) Hertz

19. c) Tuberculosis

20. c) Gravitational force

21. b) Atom

22. c) Protons and neutrons

23. a) 1g of H_2

24. b) 8

25. a) Cellulose

26. c) Mitochondria

27. b) Sun

28. a) Cultivation of fish

29. c) Mechanical energy

30. b) Velocity

31. a) 9.8 m/s^2

32. b) Communicable diseases

33. c) Tuberculosis

34. c) Gravitational force

35. c) Muscle tissue

36. b) Protista

37. c) Crop rotation

38. c) Carbon dioxide

39. a) Newton

40. c) Displacement

41. a) Force × Distance

42. b) Atom

43. c) Protons and neutrons

44. b) 8

45. a) Cellulose

46. b) Sun

47. a) 9.8 m/s^2

48. c) Mechanical energy

49. b) Communicable diseases

50. b) Velocity

51. b) Protista

52. b) Atom

53. a) 1g of H_2

54. c) Muscle tissue

55. d) Vascular tissues

56. c) Protons and neutrons

57. a) Cellulose

58. a) Mangifera indica

59. b) Atom

60. a) Force × Distance

61. a) Cultivation of fish

62. a) Newton

63. b) Velocity

64. c) Displacement

65. a) 9.8 m/s^2

66. b) Protista

67. c) Carbon dioxide

68. b) Communicable diseases

69. c) Tuberculosis

70. b) Atom

71. c) Gravitational force

72. b) Sun

73. c) Mechanical energy

74. a) Hertz

75. c) Gold

76. b) Velocity

77. a) Newton

78. a) Sublimation

79. a) Saturated solution

80. b) Lavoisier

81. a) Protons

82. c) Cell

83. b) Xylem

84. b) Whittaker

85. b) Meter/second

86. c) Gravitational force

87. a) Gravitational acceleration

88. b) Position

89. b) Frequency

90. c) Influenza

91. c) Solar energy

92. a) Nitrogen

93. c) Decantation

94. c) Molecule

95. a) J.J. Thomson

96. b) Chloroplast

97. a) Epithelial tissue

98. c) Linnaean

99. b) Rectilinear motion

100. a) Inertia

101. a) $F = G(m1m2/r^2)$

102. b) Energy

103. b) Loudness

104. a) Magnitude and direction

105. c) Catalyst

106. a) Bohr model

107. c) Platelets

108. a) The variety of species

109. c) Displacement

110. b) Resist changes in motion

111. a) Gravity acting on it

112. a) Mass and velocity

113. a) Sound waves

114. a) Mosquitoes

115. a) Evaporation, condensation, and precipitation

116. b) Phosphorus

117. a) Protons and neutrons

118. b) Rusting of iron

119. b) Ion

120. b) Potential energy

121. b) Nucleus

122. b) Nervous tissue

123. b) Photosynthesis

124. b) Potential energy

125. d) Steel

126. b) Water

127. c) Strong nuclear force

128. a) Protein synthesis

129. b) Osmosis

130. c) Friction

131. b) Oxygen

132. b) Covalent bond

133. b) 7

134. c) Combustion of methane

135. b) Ionization energy

136. a) Pascal

137. c) Mitochondria

138. b) Ductility

139. b) Gregor Mendel

140. b) H_2SO_4

141. a) Length

142. d) Carbon dioxide

143. c) Bohr model

144. b) Fight infections

145. a) Photolysis

146. c) Natural gas

147. c) Amino acid

148. b) Argon

149. b) Cerebellum

150. c) Neutralization

Printed at Repro India Ltd.